Pharmacist Podcaster

A PODCAST PLANNING GUIDE FOR PHARMACY PROFESSIONALS

Kim Newlove, RPh

©2025 Kim Newlove The Pharmacist's Voice®, LLC | www.thepharmacistsvoice.com

Copyright ©2025 Kim Newlove, The Pharmacist's Voice ®, LLC

Edited and designed by: Publishing in Doses, LLC

Print ISBN: 979-8991852906

eBook ISBN: 979-8991852913

www.thepharmacistsvoice.com

To comment on this book, send an email to the author at kim@thepharmacistsvoice.com.

All rights reserved. No part of this publication may be reproduced, stored in a retrieval system, or transmitted in any form or by any means, electronic, mechanical, photocopying, recording, or otherwise, without prior written permission of the author.

Notice

The author has made every effort to ensure the accuracy and completeness of the information presented in this book. However, the author cannot be held responsible for the continued currency of the information, any inadvertent errors, or the application of this information to practice. Therefore, the author shall have no liability to any person or entity with regard to claims, loss, or damage caused or alleged to be caused, directly or indirectly, by the use of the information contained herein.

Table of Contents

Introduction 4

Chapter 1.   **Your *Why*** 6

         Worksheet for Chapter 1 10

Chapter 2.   **How to Start** 11

         Worksheet for Chapter 2 16

Chapter 3.   **Time to Publish** 18

         Worksheet for Chapter 3 21

Chapter 4.   **Tools and Gear** 22

         Worksheet for Chapter 4 26

Chapter 5.   **MP3 Files** 27

         Worksheet for Chapter 5 30

Chapter 6.   **Research** 31

         Worksheet for Chapter 6 34

Chapter 7.   **Writing** 35

         Worksheet for Chapter 7 37

Chapter 8.   **Artwork** 38

         Worksheet for Chapter 8 42

Chapter 9.   **Media Hosts** 43

         Worksheet for Chapter 9 45

Chapter 10.   **Websites** 46

         Worksheet for Chapter 10 49

Chapter 11.   **Advertising** 50

         Worksheet for Chapter 11 55

Chapter 12.   **Your Recipe** 56

         Worksheet for Chapter 12 58

About the Author 59

Introduction

Welcome to *Pharmacist Podcaster: A Podcast Planning Guide for Pharmacy Professionals*! If you want to start a podcast, you're in the right place. This book will help you plan before you launch.

There is an overwhelming amount of information available about podcasting. From printed books and audiobooks to podcasts about podcasting, websites, coaches, online courses, YouTube videos, and even more choices, it's hard to know where to start. I felt overwhelmed by information when I started my podcast, too.

When you learn anything new, the messenger is essential. I am a pharmacist. If you're a pharmacist, too, we may think alike. I'm also an experienced podcaster. I'll tell you what you need to know so that you can avoid time-consuming rabbit holes, stuff you don't need to buy, and frustration with the podcast-planning process.

This book covers 12 topics.

1. Your *Why*
2. How to Start
3. Time to Publish
4. Tools and Gear
5. MP3 Files
6. Research
7. Writing
8. Artwork
9. Media Hosts
10. Websites
11. Advertising
12. Your Recipe

The worksheet at the end of each chapter contains critical-thinking questions, assessments, or problem-solving activities. Thoughtfully complete each worksheet. When you finish the book, you will have a plan for your podcast.

Mindset is important. Keep an open mind as you read this book. Podcasting is a creative endeavor, and there is more than one way to create a podcast. Create a recipe, not a statue. Your podcast should have ingredients that can be changed over time. It's easy to change a recipe. It's hard to change a statue. My podcasting coach, Dave Jackson from The School of Podcasting, introduced the recipe analogy to me, and I love it. I'm a pharmacist, and pharmacists use recipes. Remember to answer the questions or complete the assessments and problem-solving activities. They will help you build your podcast recipe.

As with podcast show notes, some of the products and websites mentioned in this book have links. Click them to learn more.

Happy podcasting!

CHAPTER 1

Your *Why*

Why do you want to start a podcast? People start podcasts for various reasons. Some want to be thought leaders, while others want to have fun and stay connected to friends. There are many other reasons to start a podcast. See below for more examples:

- Branding
- Build an audience
- Show or gain empathy
- Establish relationships with guests
- Market products or services
- Review products or services
- Satisfy a curiosity
- Solve a problem

Your *Why* is very important. As you start your podcast recipe, discover your *Why*. Your *Why* will guide your decisions as you plan your podcast. If you haven't figured out your *Why* yet, the following questions will help.

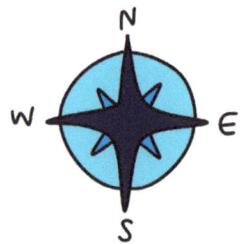

- Why do you want to start a podcast? (What's the goal?)
- Who is your audience?
- What's in it for them?
- What do you want to talk about?
- What does your audience need to hear from you?

Chapter 1: Your Why

- Once your audience hears your message, what do you want them to do with the information?
- What's in it for you to have a podcast?
- How does a podcast fit into your personal mission or your company's mission?

What's my *Why*, and how did I figure it out? *The Pharmacist's Voice Podcast* is a business tool. I use it for branding, marketing, networking, and relationship-building. Research and reflection helped me pinpoint my *Why*. Once I researched reasons that people start podcasts, I found that using my podcast as a business tool resonated with me. Your *Why* may be different from my *Why*.

I apply my *Why* to each episode I create. Let's use *episode 190 of The Pharmacist's Voice Podcast* as an example. The title is *5 Resources for pharmacist authors who want to narrate their own audiobook*. I used the episode for branding, marketing, networking, and relationship-building.

The audience is pharmacist authors who want to narrate their own audiobook. In this episode, my audience hears about resources that will help them narrate their audiobook. I flatten the learning curve, save them time, and help them feel less overwhelmed with the task. By talking about my experience with these resources, I strengthened my brand as a voice actor who narrates audiobooks. I also marketed my consulting service to my audience to position me as a resource. Hiring me will save clients time and provide the support they need. As far as networking, when I shared *episode 190* on social media, my social media followers engaged with the posts across five platforms, and I gained new followers, too. I also deepened my relationship with my audience. I know how intimidating narrating that first audiobook can be, but I also know how to get the job done. Combining empathy and authority builds trust, which is important when building relationships. In *episode 190*, I built trust with pharmacist authors in my audience who want to narrate their own audiobooks.

When we laugh, we learn. Therefore, throughout this book, *The Pharmacy Memes Podcast* will be our fictitious podcast example. In the context of this book, *The Pharmacy Memes Podcast* has not launched yet and will be hosted by long-time friends/large-chain retail pharmacists Jenny and Bill. They graduated from The University of Wyoming School of Pharmacy in 2012. They want to keep in touch using a podcast. Jenny lives in Colorado, and Bill lives in New York City. Life is busy, but they

Chapter 1: Your Why

want a reason to talk to one another at least once a month instead of just around the holidays or their birthdays. Any resemblance to real-life figures, characters, or podcasts is purely coincidental.

Jenny and Bill are planning their podcast as they read the content in this book. The answers to the question set from earlier in this chapter are below.

Q: Why do Bill and Jenny want to start *The Pharmacy Memes Podcast*? What's the goal?
A: To have fun, stay connected, and share memes they enjoy.

Q: Who is the audience?
A: The audience is retail pharmacists; however, this podcast will likely attract other meme-loving pharmacy professionals, like students, technicians, and pharmacists outside the retail setting.

Q: What's in it for the audience?
A: Edutainment, laughter, trending topics, feeling seen, and a groan when the irony hits home.

Q: What do Jenny and Bill want to talk about?
A: The humor and the story behind one pharmacy meme per month. Each episode will feature current topics, their shared experiences, and/or meme nominations from their audience.

Q: What does the audience need to hear from Jenny and Bill?
A: Being a retail pharmacist is a tough job in the year 2024. The audience needs to laugh about the stuff that usually makes them cry. They want to hear something that will chase their troubles away for a while.

Q: What do Jenny and Bill want their audience to do with the information?
A: Feel uplifted, seen, heard, and understood, and avoid quitting on a bad day.

Q: What's in it for Bill and Jenny to have this podcast?
A: Bill and Jenny want to talk more often, laugh, and have fun. They want to flip the script on their usual conversations, which are only a few times/year and usually involve venting about work and life. By searching through memes for each episode and each presenting two to one another once a month, they will talk on a regular basis, laugh, and have fun. Someday, Jenny and Bill just might publish an illustrated book about their podcast, too!

Chapter 1: Your Why

Q: How does the podcast fit into your personal mission or your company's mission?

A: *The Pharmacy Memes Podcast* is not a business, but it could become one someday. For the sake of argument, it fits into Jenny and Bill's personal mission because when they took the Oath of a Pharmacist on graduation day, they took the first line seriously about their primary concerns being, "The welfare of humanity and relief of suffering." This podcast is helping them uphold their oath. Laughter is the best medicine. They are helping their fellow pharmacists.

That ends the question set for *The Pharmacy Memes Podcast*. Remember to start your podcast recipe with your *Why*. It will be your guide.

Complete the worksheet for Chapter 1. It will help you plan your podcast.

WORKSHEET FOR CHAPTER 1

Your *Why*

Why do you want to start a podcast? (What's the goal?)

1. Who is your audience?

2. What's in it for the audience?

3. What do you want to talk about?

4. What does your audience need to hear from you?

5. Once your audience hears your message, what do you want them to do with the information?

6. What's in it for you to have a podcast?

7. How does a podcast fit into your personal mission or your company's mission?

©2025 Kim Newlove The Pharmacist's Voice®, LLC | www.thepharmacistsvoice.com

CHAPTER 2

How to Start

Find resources, listen to people, make decisions, and take action. That's how you start a podcast.

Have you ever seen maps in the comic strip *The Family Circus*? In each map, one of the four children featured in the comic wanders from point A to point B. Let's say the oldest child, Billy (around eight years old), hops off the school bus. Then, he takes *the long way* home. On the cartoon map, Billy pets a stray dog, climbs a tree, peers down a manhole, jumps rope with some kids playing in the street, splashes in his little brother's wading pool, and finally goes into his house.

There's a lot to know about podcasting, and there will be distractions when you get started. You might feel like Billy going from the bus stop to his house. If you want to get from idea to published with fewer distractions, find resources, listen to people, make decisions, and then take action.

Chapter 2: How to Start

Find resources.

Consider how you like to learn. Do you learn through reading books? Maybe you prefer to Google things and then watch YouTube videos? Audiobooks, online courses, podcasts, Facebook groups, or working one-on-one with a coach may be your jam. Find resources that suit your learning style.

Find people you trust, and listen to them.

For as much as I know about podcasting, I'm not the only person you should listen to about it. Six people I trust and listen to about podcasting are Dave Jackson, Kristen Meinzer, David Hooper, Erik K. Johnson, Steve Stewart, and Pat Flynn. I recommend you check out my list of six people first. Then, search for other people as needed. Avoid rabbit holes. Make a list of people you trust about podcasting, and then listen to them. Input from others is important when starting a podcast. I recommend you listen to more than one person.

Make decisions.

Chip and Dan Heath wrote a book titled, *Decisive: How to Make Better Choices in Life and Work*. In the book, the authors talk about how to stop the cycle of agonizing over decisions. I love this book (and audiobook)! Don't agonize over your podcasting decisions. Focus on asking yourself high-quality questions first. Then, be decisive and take action.

Once you have your *Why*, answer these high-quality questions.

1. What is your podcast called?
2. What format will you use?
3. What's the podcast about?
4. If you have guests, what criteria will you follow for inviting guests?
5. Who is on your guest list?
6. What will you ask your guests?
7. How often will you publish?
8. How will you organize topics, your guest list, and your production schedule?

9. How long will your episodes be?
10. What tools and gear do you need?
11. Where will you buy your tools and gear?
12. How will you learn how to use your tools and gear?
13. How much time do you have available to invest in learning how to use your tools and gear?
14. How will you make your MP3 files and/or videos?
15. What research and/or reflection is needed for each episode?
16. What do you need to write for each episode?
17. What will you use for artwork?
18. Which media host will you use?
19. How will you create your [podcast] website?
20. How will you advertise episodes?
21. How will you keep up with changes in the podcasting industry?

For the sake of our fictitious podcast example, *The Pharmacy Memes Podcast*, Jenny and Bill have made the following choices:

Their podcast will be called *The Pharmacy Memes Podcast*. It will have a co-host format. The podcast is about humorous pharmacy memes and the meaning behind them. They do not plan to have guests, but they will allow their audience to nominate memes for future episodes. Since they have no guests, there is no guest list, list of questions for guests, or criteria to follow for inviting guests.

The Pharmacy Memes Podcast will be published on the first day of each month. To organize topics and their production schedule, they will use Google Drive and other Google tools. Episodes will be less than 15 minutes in length due to personal bias. Their favorite podcasts are all 15 minutes or less, they are busy, and they don't want to spend a lot of time on their podcast.

Chapter 2: How to Start

Jenny and Bill have no idea what tools and gear they need, where to buy, or how to use anything. However, they want to sound good and use whatever will get them up and running quickly and easily. They each only have about 30-60 minutes per week to spend on podcasting.

Jenny and Bill are stumped about MP3 production. Neither has ever done it, and they are both intimidated by THAT kind of technology. Initially, they will not be publishing YouTube videos.

Research and reflection for *The Pharmacy Memes Podcast* will require both individual effort and collaboration. Jenny and Bill will each nominate two memes/month using a Google Drive folder. As the podcast grows, they will also consider memes nominated by email from their audience.

Agreeing on the winning meme and the meaning behind it will be as easy as a phone call, a text, or a Zoom call. They will both do some research and type their thoughts into a shared Google Doc. Then, they will schedule a call to discuss the winning meme, their research, and their experiences as part of a recorded call, which will become the podcast episode recording.

Writing will be a shared task. They agree to use a checklist on a shared Google Sheet every month.

For artwork, they plan to get permission to use the meme-of-the-month in their episode artwork. As far as creating the artwork, they have zero graphic design experience, and neither of them has access to a graphic design tool.

Jenny and Bill had no idea that a media host was needed for podcasting. They have no idea how that works.

There is no website for *The Pharmacy Memes Podcast*, and Jenny and Bill think they don't need one. People can just look for their podcast in podcast directories or podcast player apps, right?

Advertising is not something Jenny and Bill want to do. Promoting their podcast feels sales-y, so they're just going to put the podcast out there and hope people find it.

As far as keeping up with changes in the podcasting industry, Jenny and Bill don't want another job. So, they're just going to create this podcast for fun and just deal with whatever comes up when the time comes. They plan to Google stuff PRN.

Chapter 2: How to Start

That ends our update on *The Pharmacy Memes Podcast.* Sounds like Bill and Jenny have some answers, but they're nowhere near ready to launch episode one.

Jenny and Bill need to find resources, listen to people, make decisions, and take action. As we work our way through the remaining chapters, we will fill in the gaps in Jenny and Bill's plan.

Complete the worksheet for Chapter 2. It will help you plan your podcast.

WORKSHEET FOR CHAPTER 2

How to Start

**Find resources, listen to people, make decisions, and take action.
Once you have your *Why*, answer these high-quality questions.**

1. How do you like to learn?

2. Which experts will you listen to about podcasting?

3. What is your podcast called?

4. What format will you use?

5. What's the podcast about?

6. If you have guests, what criteria will you follow for inviting guests?

7. Who is on your guest list?

8. What will you ask your guests?

9. How often will you publish?

10. How will you organize topics, your guest list, and your production schedule?

11. How long will your episodes be?

12. What tools and gear do you need?

13. Where will you buy your tools and gear?

14. How will you learn how to use your tools and gear?

15. How much time do you have available to invest in learning how to use your tools and gear?

16. What research and/or reflection is needed for each episode?

17. What do you need to write for each episode?

18. How will you make your MP3 files and/or videos?

19. What will you use for artwork?

20. Which media host will you use?

21. How will you create your [podcast] website?

22. How will you advertise episodes?

23. How will you keep up with changes in the podcasting industry?

©2025 Kim Newlove The Pharmacist's Voice®, LLC | www.thepharmacistsvoice.com

CHAPTER 3

Time to Publish

How long does it take to go from idea to published? It depends. Why does it matter? Because comparing your availability to your production time will help you understand how often to publish your podcast episodes.

My Numbers

A 15-minute solo show takes roughly 3 hours from idea to published.

A 60-minute interview show takes roughly 4.5 hours from idea to published.

Based on your personal preferences, tools, experience level, podcasting team, and a host of other factors, your times will be different from mine. If you talk to a large number of podcasters who know their numbers, you'll get a good idea about how long it might take you to publish an episode.

Speed comes with time and experience. For a first-time, do-it-yourself podcaster, episode one will take longer to put together than episode 20.

To hear several podcasters talk about their time from idea to published, listen to *episode 803 of The School of Podcasting Podcast*. In this episode, the host - Dave Jackson, crowdsourced input from more than a dozen do-it-yourself podcasters about their time from idea to published. I also created an episode about the topic. Listen to *episode 126 of The Pharmacist's Voice Podcast*. To get more opinions, talk to more podcasters.

Chapter 3: Time to Publish

Why 3 hours for my 15-minute solo show? If the final length of a solo show is only 15 minutes, you may be wondering, "What happens during the other 2 hours and 45 minutes?" Among other things, I spend time doing research, writing, editing audio, creating artwork, uploading everything to my media host, and advertising. I don't spend 3 hours all in one sitting. In fact, some episodes are put together over a few weeks. It all starts with an idea. That's why the first sentence of this chapter includes the phrase, "From idea to published."

Why 4.5 hours for my 60-minute interview show? If the final length of an interview show is only 60 minutes, what happens during the other 3.5 hours? For one thing, that 60-minute interview show might contain a 2-3 minute introduction and a 2-3 minute outro (conclusion), which needs to be recorded and edited separately.

Recording and editing interviews is time-consuming. I have to invite the guest, send a remote-recording program link (or set up an in-person interview), show up at the scheduled time, chat in a friendly way before recording, record the interview, have some sidebar conversations that will be edited out, and have a brief send-off conversation after we wrap up our interview. I have to listen to the entire interview again and edit it. A 55-minute interview may start off as a 65-minute raw recording.

But wait, there's more! I spend time researching guests so I can choose interview questions that serve my audience and satisfy my Why. I also write a title, craft a written description for the podcast players, write show notes for the blog post on my website, create artwork, upload everything to my media host, and advertise. It's a time-consuming process, and I don't do it all in one day.

Whether you publish monthly, daily, or something in between, publishing on a regular schedule will build trust with your audience. Great content is a must, but a regular schedule is important too.

How did I pick my production schedule? I chose a weekly schedule arbitrarily in January 2020, and I was able to keep up with it. So, *The Pharmacist's Voice Podcast* has been on a weekly schedule for more than four years. Audio production is a big part of my business, and podcasting is a priority. I spend about 3 hours/week on podcast-related activities for my business. If I need to change my production schedule at some point, I will.

Chapter 3: Time to Publish

Bill and Jenny plan to publish *The Pharmacy Memes Podcast* on a monthly basis. How much time do they have available compared to how long it will take them to publish monthly? They both have a maximum of 4 hours/month to collaborate on this podcast. Their instinct to choose a monthly schedule seems logical. They haven't produced episode one yet, so we'll see how they fare in the coming chapters.

Practice podcast episodes improve workflow and shorten production time. Any runner who has done an 8-week *Couch to 5K Training Program* knows that you can't train for a 5K race in one day. Each training session builds endurance and speed. Race fans don't show up to watch training sessions, and your audience will never know that you created practice episodes. You can redo, trash, or publish your practice episodes. The point of practicing is to improve your workflow and shorten your production time so that you can find a production schedule that works for your availability.

Remember that podcasting is a creative activity. You may need to trial-and-error your way into a workflow and production schedule that works for you. The biggest takeaway of the chapter is that comparing your availability to your production time will help you understand how often to publish.

Complete the worksheet for Chapter 3. It will help you plan your podcast.

WORKSHEET FOR CHAPTER 3

Time to Publish

1. Time assessment: How much time do you have available to publish podcast episodes?

2. On average, how much time does it take to create a podcast episode?

3. Compare your available time to how long it takes to create an episode. How often will you be able to publish?

4. Do you need to add someone to your team to free up your time?

©2025 Kim Newlove The Pharmacist's Voice®, LLC | www.thepharmacistsvoice.com

CHAPTER 4

Tools and Gear

Tools and gear are important ingredients in your podcast recipe. In this chapter, we'll talk about what you need, where to shop, how to learn how to use everything, and how long it takes to get up and running.

What do you need?

You need to sound great and create MP3 files. At a minimum, you need a microphone and recording software. Want to get started for $100 or less? Here's a list of 6 things you need, some of which you already have.

1. The computer you already own ($0)
2. [A Samson Q2U USB/XLR microphone](#) (Some accessories are included.) ($69.99, April 2024)
3. A USB to USB-C adapter if needed (to connect the mic to the computer) ($8.99, April 2024)
4. Headphones or earbuds that work with your computer ($0)
5. A digital audio workstation (DAW)/recording software, such as Audacity ($0) or GarageBand ($0). *Make sure your computer's operating system works with the DAW (recording software) you choose.
6. Remote recording software ($0, Zoom free version)

Chapter 4: Tools and Gear

If you enjoy podcasting and stick with it, you can upgrade down the road. For example, new tools and gear could be a podcast anniversary gift to yourself.

Where to shop?

Three places I buy tools and gear are Sweetwater.com, Amazon.com, and Guitar Center. Sweetwater is my go-to because their online store carries everything I need, and their helpful customer service is staffed by humans. If you like to "buy local," Guitar Center is a good choice; you can try equipment in-store and ask questions when you visit. Convenience and speed win customers over sometimes, and Amazon is a convenient option for buying tools and gear quickly and easily. I have bought from all three places. Buy from whomever you like.

How will you learn how to use everything?

Learning anything new can be a challenge. Consider how you like to learn and approach learning how to use your tools and gear the same way. Do you like reading user manuals? Trial and error? YouTube videos? Online courses? Private coaching? Facebook groups? There are plenty of options, and you need to be decisive because you need to sound great and generate MP3 files so you can launch your podcast.

When I started in the voiceover industry in 2017, I knew nothing about using microphones or creating audio files. I felt frustrated. I didn't even know how to hook up a microphone so that my computer recognized it as my audio input! By the time I started podcasting in 2019, however, I had mastered the two major tasks you must now learn: how to use your microphone and how to create MP3 files.

How did I learn how to hook up a microphone and use my first DAW (Audacity)? My colleague Jonah Rosenthal helped me. We had a Zoom lesson for the mic, and I bought his 2-hour online course to learn Audacity.

When I began audiobook narration training in 2019, I switched from Audacity to a more robust DAW called Studio One Artist. It had the features I needed for the audiobook narration part of my business.

Chapter 4: Tools and Gear

The switch from Audacity to Studio One Artist was not easy for me. It was like learning how to fly a rocket ship across the street instead of just walking across. It seemed like overkill, and I cried when I learned how to use it. I don't cry much, so this was a big deal. I ended up getting a mentor named Don Baarns, who taught an online course about Studio One Artist, gave me some helpful macros (shortcuts), and welcomed me into a helpful Facebook Community with thousands of members who also use Studio One Artist. On rare occasions, I have also paid for private coaching with Don. In review, I found a mentor, took an online course, joined a Facebook group, paid for private coaching, and practiced. My story is not meant to scare you. Rather, I want to keep it real about how I learned. You may find it challenging and need help, too. You don't have to do everything alone. It's okay to get help.

How long before you're up and running?

You will be up and running as soon as you learn how to use your tools and gear. How long will that take? It depends. Looking back at my experience with Studio One Artist, I knew how to use it well enough to produce podcast episodes in less than one month. Depending on how much time and effort you put into learning how to use your tools and gear, you may be up and running sooner or later.

Set a goal.

Set a goal to learn how to use your tools and gear in 30 days or less. Buying tools and gear and learning how to use everything does not need to be a long, drawn-out process. Once you know how to at least press start and stop, set your [start and end] markers, and mix down an MP3 to meet your media host's specifications, you're one step closer to publishing podcast episodes!

In the introduction, I mentioned that there is an overwhelming amount of information available about podcasting. Some new podcasters get stuck in the weeds, researching which microphone to buy. Don't let that happen to you. A Samson Q2U is a great starting point. Learn how to use it well, and you'll be in good shape. When it comes to tools and gear, learning how to generate MP3 files is your biggest concern. The next chapter is all about MP3s.

Chapter 4: Tools and Gear

Let's check in on Bill and Jenny from *The Pharmacy Memes Podcast.* They are each buying a Samson Q2U microphone from Sweetwater.com. Bill is a Mac user. He will use his free Zoom account to record remotely and edit using GarageBand. He plans to join a free Facebook group about GarageBand and either watch YouTube videos or use recommendations from group members to learn how to make MP3 files. Good plan, Bill and Jenny! We'll check in on them in Chapter 5.

In this chapter, you learned which tools and gear you need, how to get started for under $100, where to shop, that it will take time to learn how to use everything, and that you need to set a time-bound goal for getting up and running.

Complete the worksheet for Chapter 4. It will help you plan your podcast.

WORKSHEET FOR CHAPTER 4

Tools and Gear

1. What do you need to buy?

2. Where will you shop for tools and gear?

3. How will you learn how to use everything?

4. How long do you think it will take you to get up and running?

©2025 Kim Newlove The Pharmacist's Voice®, LLC | www.thepharmacistsvoice.com

CHAPTER 5

MP3 Files

MP3 files are essential to your podcast recipe. An MP3 file is a digital file that contains audio information. Much like a JPEG is synonymous with a digital picture, the term MP3 is synonymous with an audio file. If you listen to music, podcasts, or audiobooks on a smartphone, tablet, or computer, you listen to MP3s.

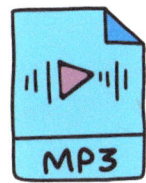 

You can't have an audio podcast without an audio file. Media hosts distribute audio files to podcast players. Ultimately, your audience will listen to your MP3 files on their podcast players.

How do you make an MP3? Below is a list of four ways to generate MP3 files for a podcast, followed by a brief explanation of each one. There are other ways; these are just four examples.

1. Do it yourself.
2. Hire an editor.
3. Hire a concierge podcasting service.
4. Join a podcast network that will generate MP3s for you.

Chapter 5: MP3 Files

Do it yourself.

Your DAW will generate your MP3 files. In *episode 210 of The Pharmacist's Voice Podcast*, I talked about 20 audio engineering skills I use. If you're not familiar with audio editing terminology, check out that episode. With time and practice, you will be able to generate your own MP3 files.

Hire a podcast editor.

A podcast editor can generate your MP3 files for you. In *episode 270 of The Pharmacist's Voice Podcast*, I interviewed a professional podcast editor named Steve Stewart. The two biggest takeaways were the benefits of hiring a podcast editor and that Steve can help you find one. To learn more, read the *show notes for episode 270 of The Pharmacist's Voice Podcast.*

Hire a concierge podcasting service.

A concierge podcasting service can generate your MP3 files for you. But, wait…there's more! They can also make it possible for you to "step up to the mic" and just talk. They use their recording software to capture your voice. Then, they edit everything to your liking and generate your MP3 files. If you want, they can also upload each MP3 to your media host. If you want a full-service experience, hire a concierge podcasting service. In *episode 255 of The Pharmacist's Voice Podcast*, I interviewed Audivita Studios Founder and CEO David Wolf. *Book a discovery call with David*, and see how Audivita can help you with your podcast.

Join a podcast network that will generate MP3s for you.

A podcast network may be able to help you generate your MP3 files. I can't speak on behalf of any network; however, I have spoken to other podcasters whose MP3s are produced by their network. When weighing the pros and cons of joining a podcast network, factor in MP3 production.

That concludes my list of four ways to generate MP3 files for a podcast. Next, we'll see which of the four options Jenny and Bill from *The Pharmacy Memes Podcast* chose.

Jenny and Bill are super excited to hear that there is more than one option for generating MP3 files! They were starting to feel backed into a corner with all the new responsibilities of publishing a podcast. They have chosen the podcast editor route. After listening to *episode 270 of The Pharmacist's Voice Podcast*, they used the Google Form mentioned by Steve, found five editors, picked one, and have him on stand-by waiting to edit episode one. Woot! It feels like a weight has been lifted! An experienced editor joined their team!

In this chapter, you learned what an MP3 file is, why you need one, and four ways to generate them. If you want to have an audio podcast, you need to create audio files. If you no longer want to create an audio podcast, it's okay to change your mind. Consider starting a YouTube channel (video podcast). Although I syndicate my audio podcast to YouTube, neither of my two podcasts are video podcasts. I just produce the occasional YouTube video. I can't help you with video podcasts, but a concierge podcast editor like *Audivita* or *Polish Your Business* can.

Once you decide if you're moving forward with an audio podcast, complete the worksheet for Chapter 5. It will help you plan your podcast.

WORKSHEET FOR CHAPTER 5

MP3 Files

1. Which DAW (recording software) will you use?

2. How will you create your MP3 files? (DIY, editor, concierge service, network, or other?)

©2025 Kim Newlove The Pharmacist's Voice®, LLC | www.thepharmacistsvoice.com

CHAPTER 6

Research

Research and writing go together like peanut butter and jelly. Both will help you get organized and shorten production time. We'll start with research in this chapter and move on to writing in Chapter 7.

For the context of this book, research is anything that helps you develop a podcast episode. From looking up words, facts, and attributions to reflecting on your personal experiences and watching a future guest's YouTube channel, research is a broad term. The best practice here is to complete your research before you press "Record." More prep = less editing and less overall production time.

Two of my favorite resources are *Merriam-Webster's Online Dictionary and Thesaurus* and *Google*. Stumbling over words during a recording frustrates me. Thinking about what I'm going to say and writing a well-planned outline before recording helps. Google helps me research websites and social media handles that end up in the podcast show notes, too.

Research for interviews is straightforward. I like to learn about my guests before I interview them. I often listen to their most recent interviews on other podcasts or YouTube channels so that my interview is different. Looking at their recent social media posts, reading their blog, or listening to their audiobook also helps me craft questions that I think my audience wants answered. Research helps me prepare for interviews.

Chapter 6: Research

Jenny and Bill want to record a practice episode of *The Pharmacy Memes Podcast*. They each brought two memes to the table. Then, they picked their favorite one. It features Mufasa and Simba from Disney's *The Lion King*. Simba asks, "Dad, what's a weekend?" And Mufasa replies, "I don't know, son. We're pharmacists."

When they push "Record," Jenny and Bill talk about the meme and the meaning behind it. Things go really well for about 3 minutes. Then, things spiral downward quickly. Their practice podcast turns into a cathartic release instead of helping them and their audience laugh about something that usually makes them cry. As large-chain retail pharmacists, Jenny and Bill are expected to work every other weekend. They look forward to their weekends off, but they often trade at least one of their weekend days so that they can attend one of their kids' activities, help their aging parents, or take a child or parent to an appointment during the day on a weekday. The struggle is real. They both go on a power rant about how so many people call off on the weekends that the scheduler can't get all the shifts covered, and they are forced to come to work because they hold management roles.

This practice episode is a real downer. After 12 minutes, they haven't laughed much, and they're not really having fun. Jenny and Bill started the podcast to have fun, stay connected, and talk about memes they enjoy. Before they pressed record, they didn't brainstorm what might uplift them or their audience or what might help anyone feel seen or understood. They stop the recording. This is not what they had in mind. After all, if laughter is the best medicine, this episode is poison.

After some discussion, Jenny and Bill decide that they need to scrap the 12-minute practice episode and try again. This time, Jenny and Bill record 13 minutes of content. It's only marginally better. They're not sure if they want to scrap it, redo it, or publish it. They wait to decide. Find out more about their second practice episode in the next chapter.

When I started my first podcast, I didn't have a workflow. Sometimes, I would stop a recording to look something up. I either had to listen to where I stopped and pick up from there or scrap the whole thing and start over. It didn't take me long to learn that researching facts and guests, collecting my thoughts, and writing outlines and questions saved time and frustration. Plus, I would communicate more effectively with my audience.

Chapter 6: Research

Research and writing are now part of my podcasting workflow. For example, in *episode 265 of The Pharmacist's Voice Podcast*, I talked about how to pronounce omeprazole and Prilosec. I researched the pronunciations and reflected on how I struggled to say the brand name when I was in pharmacy school. As a pharmacy intern, I said "PRILL-o-sec" instead of "PRY-lo-sec." I got corrected by other pharmacy staff until I got it right, and it was hard for me to break the habit of saying it wrong. For the omeprazole episode, I created a chronological outline so I could start with the introduction, deliver my main message, and wrap things up with a conclusion, which I refer to as an "outro." Because I prepared well, I edited less, and my production time was the shortest it had ever been. Once research and writing become part of your workflow, your production time will be shorter, too.

In addition to saving time and frustration, research will also help you stay true to your *Why* and be the podcast your audience needs. In my example, no one wants to say drug names wrong, get corrected by their colleagues, or struggle to break a bad habit. Episode 265 helped my audience say omeprazole and Prilosec correctly, but it also further branded me as an authority on drug name pronunciations and as a medical narrator. I showed empathy and authority, which builds trust. Therefore, this episode also deepened my relationship with my audience. They trust me when it comes to drug name pronunciations.

Do research and get organized before you press "Record." You will save time, avoid frustration, improve audience satisfaction, and satisfy your *Why*.

Complete the worksheet for Chapter 6. It will help you plan your podcast.

WORKSHEET FOR CHAPTER 6

Research

1. How will you research topics?

2. How will you research guests?

3. How will you pick interview questions?

4. How is your first episode related to your *Why*?

©2025 Kim Newlove The Pharmacist's Voice®, LLC | www.thepharmacistsvoice.com

CHAPTER 7

Writing

Writing will also help you get organized and shorten production time. You will write two types of content:

1. Public-facing content that everyone sees
2. Private content that only the podcaster sees

When I was growing up in the '80s, one of my favorite superheroes was Superman. It blew my 8-year-old mind that a superhero could be a nerdy-looking guy like Clark Kent one minute and Superman the next. Kryptonite was the icing on the cake. The fact that Superman had a weakness was fascinating.

At a minimum, the public-facing/Clark Kent-type content includes the episode title, number, description for the podcast players, and show notes for the podcast's website. The written content that will save the day, like Superman, is private content that only the podcaster sees. Examples include an outline for each solo show, an intro and outro for each episode, questions for each interview, a production schedule, a topics list, a guest list (and waitlist for guests), a calendar link for scheduling interviews, invitation email template, embed code for the media host, and a comprehensive episode list.

Chapter 7: Writing

Failing to plan is planning to fail. Underestimate the power of research and writing, and your message will be less powerful, like Superman holding a chunk of kryptonite.

Co-hosts Jenny and Bill at *The Pharmacy Memes Podcast* picked a new meme, and they're working on their second practice episode. This time, they picked a meme featuring a pharmacist in a white lab coat holding a sign that says, "I have no idea why your copay is $20. Call your insurance company."

Jenny and Bill are prepared this time. They typed notes into their Google Document, shared it with one another before the recording, and got on the same page. In the end, they have a list of bullet points they want to cover.

Good news! Bill and Jenny recorded 13 minutes of content. They first described the meme and the meaning behind it. Then, they shared their personal experiences with patients who didn't know their copay or their deductible. Finally, they talked about what it would sound like if they "Yes, and-ed…" their patients' comments. (They learned this strategy from Dr. Cory Jenks' book *Permission to Care: Building a Healthcare Culture That Thrives in Chaos*.) They had fun and also recorded a funny podcast episode their audience will enjoy. Good work, Jenny and Bill!

Like Jenny and Bill, when I have a plan before I record, podcasting is easier for me, too. That's why more prep = less editing and production time. Everything mentioned at the beginning of the chapter and more is part of my writing strategy. From writing helpful titles to creating outlines and interview questions, I do it all.

Staying organized can be a challenge. I like to keep as much as possible in a binder, but I also use a word-processing program called Pages, Excel spreadsheets, Canva, and more.

Your big takeaway from this chapter is that research and writing will help you get organized and save time.

Complete the worksheet for Chapter 7. It will help you plan your podcast.

WORKSHEET FOR CHAPTER 7

Writing

What do you need to write for episode one of your podcast?

1. Episode title

2. Episode description (for the podcast players)

3. Blog post (for your podcast website)

4. Outline (for your solo show)

5. Interview questions (for an interview)

6. What are the first 5 episodes on your production schedule?

7. What are the top 5 topics you will talk about (solo show)?

8. What are the top 5 names on your guest list (interview show)?

9. What are the top 5 names on your waitlist (interview show)?

©2025 Kim Newlove The Pharmacist's Voice®, LLC | www.thepharmacistsvoice.com

CHAPTER 8

Artwork

Readers judge a book by its cover, and podcast listeners judge podcasts by their artwork. Podcast artwork attracts listeners. Therefore, you need artwork for your podcast.

Terminology and specifications matter. I'll list three types of artwork your podcast may need and give a short description of each.

1. Podcast artwork
2. Episode artwork
3. YouTube thumbnails

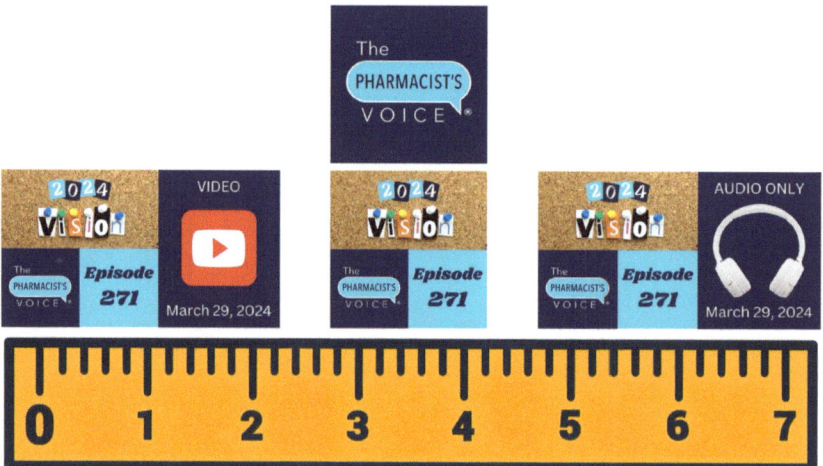

Podcast artwork

Podcast artwork is the image that appears in podcast players and podcast directories for your podcast. Your media host will provide specifications (specs) for your podcast artwork. For example, a square, 1620 X 1620-pixel JPEG under 500 kb with sRGB color space checks all the boxes for my media host, which is Libsyn. I hired a graphic designer to create my podcast artwork, and she met Libsyn's specs for me.

Episode artwork

Episode artwork is optional. It's the image that is paired with an individual podcast episode. Either make a custom design for each episode or just use your podcast artwork for each episode. Check with your media host for episode artwork specs. I create episode artwork for each episode of *The Pharmacist's Voice Podcast* using Canva. When the artwork exceeds 500 kb, I use a tool called Squoosh to make it smaller.

YouTube thumbnails

A YouTube thumbnail is artwork that is paired with each audio podcast episode that is syndicated to YouTube by a media host. You can either create artwork for each individual episode or use the same image for every episode.

Note that syndicating a podcast episode to YouTube using a media host is not the same as having a YouTube podcast. Syndicated audio episodes use a static image paired with an MP3 file. If you create the video version of an episode, you can also upload it to YouTube with a thumbnail.

Thumbnails for my audio-only episodes on YouTube feature a pair of headphones, and the video episodes feature a YouTube logo. Audio episode thumbnail specs come from your media host, and YouTube thumbnail specs come from YouTube.

Your artwork will appear in the podcast players and on YouTube when you use the right specs. I accidentally used the wrong specs for some episode artwork recently, and my podcast artwork showed up instead of the episode artwork. With a quick redo in Canva, I was able to update the artwork in Libsyn. It took some time for the podcast players to update, but the problem was eventually fixed. Learn from my mistake. Use the right specs, and you'll be fine.

The Pharmacy Memes Podcast needs artwork. Jenny and Bill do not have Canva, and they don't know how to create podcast artwork or YouTube thumbnails. Neither are very artistic, and they

Chapter 8: Artwork

have been dreading this ingredient in their podcast recipe. They don't have a media host yet, and they don't even know which specs to use.

Jenny and Bill like to use Google to learn new things. So, Jenny and Bill Google "podcast artwork specifications" and see three criteria: square, JPEG or PNG, and 3000 X 3000 pixels. They also Google, "YouTube thumbnails." YouTube specs say a thumbnail should be 1280 X 720 pixels. Now that they have the specs, they need Canva.

But first, Jenny and Bill talk about what images and text should be in the artwork. They decide to pick a humorous, cartoonish design with a silly font. They want their audience to laugh about the things that usually make them cry. They want their artwork to include something that visually represents "chasing troubles away." They also like the tagline, "Laughter is the best medicine." Bill had a thing for PAC-MAN as a kid. He liked chasing the ghosts and eating the little dots, so he proposes a PAC-MAN theme. Jenny likes it.

Next, they brainstorm how to get access to Canva. Jenny offers to ask her good friend Mandy, who is a junior high art teacher, to help. When Jenny calls Mandy, Mandy is excited to help. Together, Jenny and Mandy make the designs below.

Chapter 8: Artwork

When Bill sees the podcast artwork and YouTube thumbnail designs, he loves them! PAC-MAN is chasing something away. It looks like humans, but that's fine. The words "Pharmacy" and "Memes" are in a yellow PAC-MAN-themed font, and the tagline, "Laughter is the best medicine" is at the very bottom in black. The artwork checks all the boxes for Bill, and it costs them nothing but time to create. Mandy (the teacher) was just happy to help. All Jenny has to do is buy Mandy a coffee next time they meet up.

It's okay to start with basic artwork made with Canva and update it later if needed. After all, a podcast is a recipe, not a statue. Done is better than perfect, and it's okay to change the recipe.

Complete the worksheet for Chapter 8. It will help you plan your podcast.

WORKSHEET FOR CHAPTER 8

Artwork

1. How will you create your podcast artwork and meet specs?

2. Will you use episode artwork? If yes, then how will you create it (and meet specs)?

3. How will you create your YouTube thumbnails (and meet specs)?

4. Which tools do you need to subscribe to or learn how to use?

5. How will you find the right graphic designer for you (if you want to hire one)?

©2025 Kim Newlove The Pharmacist's Voice®, LLC | www.thepharmacistsvoice.com

CHAPTER 9
Media Hosts

How do your podcast ingredients get from your computer to your audience? Your MP3 file, written details, and artwork are distributed to podcast players via a podcast hosting platform, which I will refer to as a *media host* in this book. A media host stores or "hosts" your podcast ingredients. Three popular media hosts are Libsyn, Podbean, and Spotify for Podcasters.

Features affect pricing. I pay $20/month for Libsyn's Advanced Plan (April 2024), but other plans are also available. I started using Libsyn in 2019, and I'm a satisfied customer. Below are some reasons I use Libsyn.

- My podcasting coach (Dave Jackson from The School of Podcasting) recommended it.
- $20/month is affordable to me.
- I understand the user interface.
- The human-staffed customer support is helpful. I have used it several times.
- I like to see my download numbers and which countries are downloading episodes.
- The storage space and bandwidth meet my needs.
- Libsyn distributes my podcast to all the major podcast players and YouTube.
- They provide embed code so I can put a podcast player on my website.

Chapter 9: Media Hosts

When you shop for a media hosting service, research plan pricing, the user interface, customer service, stats, storage and bandwidth, distribution, and other features that are important to you. Beware of free media hosts! You get what you pay for. If researching media hosts online doesn't answer your questions, ask other podcasters for input. Ask which media host they use, why they chose them, and if they're happy with them.

Once you have the key ingredients for your first episode, it's time to sign up. You need the MP3 file for episode 1, written details (episode title, number, description, etc.), artwork, and release date/time. The media host will distribute everything for you at the time you schedule it to be released. It's just that easy!

Jenny and Bill are almost ready to publish their first podcast episode. They have the MP3 file for episode one, the written details, and the artwork. They want to publish on the first day of each month. They're going to sign up for a basic plan with Libsyn and set everything to release on April 1. Perfect timing! April Fools' Day is the funniest day of the year!

In review, your media host is an important part of your podcast recipe. It hosts your content and distributes your podcast to your audience. Pick a host that fits your budget and offers the features you need.

Complete the worksheet for Chapter 9. It will help you plan your podcast.

WORKSHEET FOR CHAPTER 9

Media Host

1. What's your budget for a media host?

2. What features do you need?

3. Which media host will you use?

4. When will you sign up?

5. Bonus: Which podcast directories do you want to be in, and how will you get into them?

6. Bonus: Do you need embed code to put a podcast player on your website?

©2025 Kim Newlove The Pharmacist's Voice®, LLC | www.thepharmacistsvoice.com

CHAPTER 10
Websites

Podcast websites help listeners find podcasts. Podcast directories and search engines help listeners find podcasts, too, but a podcast website provides a shortcut. Every fall, I go to an amusement park in Sandusky, Ohio, called Cedar Point. Lines for the most popular rides are over an hour long on busy days. Fast Lane Passes cost extra, but they get riders to the front of the line. If you want to give your audience a Fast Lane Pass to find your podcast, create a podcast website.

There's more than one way to put a podcast on a website. I'll give you four examples.

First, *The Pharmacist's Voice Podcast* lives on my business website. Click the Podcast tab on www.thepharmacistsvoice.com to find it. I have a business with a podcast, and putting my podcast on my business website was the right choice for me.

Second, podcasts can have free-standing websites. My second podcast (*The Perrysburg Podcast*) has a free-standing website at www.perrysburgpodcast.com. I bought a domain for it and used a website builder called Podpage. Podpage is easy to use, and I would recommend it for your podcast website.

The third example is a website created by a media host. To see an example, visit Tony Guerra PharmD's *Pharmacy Residency Podcast* website at www.pharmacy.libsyn.com.

Chapter 10: Websites

The fourth and final example is Linktree. If you're creative, you can use a free landing page, like Linktree, as an alternative podcast website. Just program links to your podcast into your Linktree and share the link. It's not really a website, but it does provide a shortcut to your podcast on the internet.

Other than providing a direct path to a podcast, a podcast website has other benefits. It can:

- Display show notes (blog posts) and links
- Offer other episodes to your audience
- Share your social media links
- Showcase your brand colors and episode artwork
- Display a podcast player for each episode so that your audience can listen to episodes directly from your website
- Make offers on your website, like joining an email list or getting a lead magnet
- Sell merchandise
- Share your background information using an "About" tab
- Provide contact information in case your audience wants to ask a question or make a comment
- Improve SEO (search engine optimization)

Podcast websites vary in cost. Depending on whether you add your podcast to an existing website, get a free-standing podcast website, use the website provided by a media host, or use Linktree, your domain, web hosting, a website builder, or a premium landing page could add up. Choose a website that fits your needs and budget.

Need ideas for your podcast website? Imitation is the sincerest form of flattery. Look at ten other podcast websites for ideas. From professionally-produced shows to small, independent ones, there are plenty of podcast websites to visit. Note what you like and don't like so you can plan how you want your podcast website to look. Always keep the audience in mind, and answer these high-quality questions as you visit those ten podcast websites:

Chapter 10: Websites

- Is the landing page easy to use on desktop *and* mobile?
- Does the site have a custom domain?
- Do you like how the blog posts, episode artwork, and podcast players are organized?
- Do they have an About page and a Contact form?

As you consider how your audience will find your podcast, keep your podcast launch date in mind. If you choose to build a podcast website, launch it the same day you launch your podcast. Your podcast website is an important ingredient in your podcast recipe, and timing is everything.

What are Jenny and Bill from *The Pharmacy Memes Podcast* doing for a website? They're going to use the free website provided by their media host. They're just starting out, and they want to keep the start-up phase simple and low-cost. The website automatically launches when the podcast does, and they feel good about their initial podcast website.

In review, your audience needs an easy way to find your podcast. I strongly recommend you create a podcast website and launch it the same day you launch your podcast.

Complete the worksheet for Chapter 10. It will help you plan your podcast.

WORKSHEET FOR CHAPTER 10

Websites

1. How will your audience find your podcast?

2. What kind of podcast website will you choose?

©2025 Kim Newlove The Pharmacist's Voice®, LLC | www.thepharmacistsvoice.com

CHAPTER 11

Advertising

Advertising a podcast is different from advertising products, services, or businesses on a podcast. In this chapter, I will talk about advertising a podcast first. Then, I'll briefly cover advertising products, services, or businesses on a podcast.

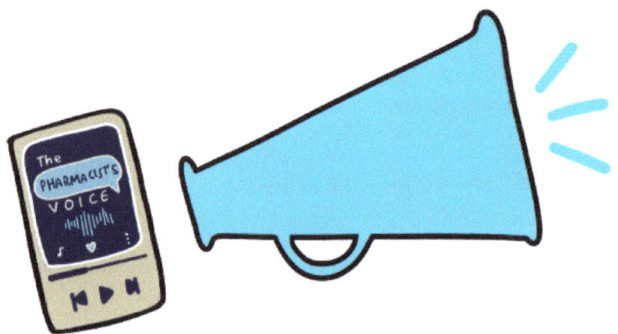

There is more than one way to advertise a podcast. The goal is to get the word out about either your podcast in general or each individual episode. Out of the dozens of ways I could advertise my podcast, I use eleven and have thought about using seven others. Below is a list of eleven ways I advertise my podcast, followed by a brief description of each one.

1. Social media
2. Pod-guesting
3. Call to action
4. Call to connect
5. QR code
6. Word-of-mouth
7. YouTube videos
8. Blog post
9. Podcast trailer
10. Newsletter
11. Email signature

Chapter 11: Advertising

1. My weekly **social media blast** advertises my podcast episodes on LinkedIn, Twitter, Facebook, and Instagram. I combine the episode artwork, written details, and a link to the episode in each social media blast.
2. **Pod-guesting** is a good deal for me as a guest. The host does the heavy lifting on the production side, my podcast gets exposure, I network with the host, and I develop a brief relationship with their audience.
3. I end podcast episodes with a **call to action**, such as, "Subscribe to or follow *The Pharmacist's Voice Podcast*."
4. Using a **call to connect** is similar to a call to action. I encourage my audience to connect using my social media links in the show notes. Once we're connected, my new episode posts become part of my network's social media feeds.
5. I use **QR codes** on business cards to advertise my podcasts.
6. **Word-of-mouth** is still a great way to advertise; I simply tell people about my podcast. When appropriate, I also invite people to follow the podcast via direct message or email.
7. I advertise my podcast during my **YouTube videos**. At a minimum, I include my podcast artwork and the website.
8. Words are "Google juice" and have "SEO power." Therefore, each podcast episode gets a **blog post** on my website. When someone does a search, my podcast might be in their results.
9. A **podcast trailer** is a short podcast episode that explains who you are, what your podcast is about, and why someone should listen to it. Length is typically 2 minutes or less. I have only done this on my second podcast, *The Perrysburg Podcast*. Episode one was the trailer.
10. An email **newsletter** can advertise podcast episodes. I started a weekly newsletter for *The Perrysburg Podcast*.
11. Podcast player icons appear in my business **email signature**. When I email someone from my business email account, I also advertise my podcast to them in the signature.

There are many more ways to advertise a podcast. Here's a list of seven I have not tried yet, followed by a brief description.

Chapter 11: Advertising

1. Trailer swap
2. Episode swap
3. Audiograms
4. Shorts
5. Paid ads
6. Dynamic ad insertion
7. Reverse engineering

1. When two podcasters share one another's trailers, it's called a **trailer swap**. Avoid confusion! There is an art to sharing a trailer so that both audiences are aware of the swap.

2. Similar to a trailer swap, an **episode swap** involves two different podcasts. The episode could be an interview, a panel discussion, a solo episode swap, etc. Avoid confusion! As with trailer swaps, there is an art to swapping episodes so that both audiences are aware of the swap at the beginning of the episode.

3. **Audiograms** use a combination of audio podcast clips and artwork to make short video advertisements for social media. A waveform image is typically included. Headliner and Wavve are popular audiogram generators.

4. Similar to audiograms, **Shorts** on YouTube and other social media platforms advertise podcast episodes using video.

5. **Paid advertising** opportunities are plentiful. Getting the podcast in front of new eyes and ears is the goal. Three examples are Facebook ads, magazine ads, and 30-second spots on podcast networks.

6. **Dynamic ad insertion** puts custom advertisements into podcast episodes without baking them into the podcast recording. Your media host may offer this feature.

7. How did you hear about your favorite podcast? **Reverse engineer** how you found out about your favorite podcast, and use that method to advertise yours.

Chapter 11: Advertising

Again, the goal of advertising a podcast is to either get the word out about your podcast in general or get the word out about each individual episode. There is more than one way to advertise a podcast, and you need to figure out what works for you and feels right.

Now, I'll briefly cover advertising products, services, or businesses on a podcast. Some people call this podcast monetization or earning profit from your podcast.

The most common question I get asked about my podcast is, "How much money do you make off your podcast?" In my mind, they're asking, "Kim, when you press 'publish,' how much money drops into your bank account?" It's hard to say how much I ultimately earn from my podcast, but I can tell you that no money ($0) is deposited in my bank account when I press Publish.

My podcast is a business tool. I use it for branding, marketing, networking, and relationship-building. My audience uses it to get to know, like, and trust me through my podcast. Abstractly, I get paid in fun, relationships, and recognition from my audience and guests.

A more concrete answer is that my audience buys products or services I talk about on my podcast. For example, they buy my online courses or they hire me for voiceover gigs and consulting.

My guests also profit from my podcast. Some examples are products, like books and online courses; services, like financial planning or career coaching; or their businesses, like medical writing or their independent pharmacies. It's impossible to assign a dollar figure to the amount of money that changes hands because of my podcast.

Podcasting is "the long game" for me. Something I published three years ago might pay off this year. I work in the "gig economy," and I have a lot of different income streams. There is no payout per episode. Even if there were, every podcast is different, and what works for me may not work for you.

To learn more about monetizing a podcast, read the book *Profit From Your Podcast* by Dave Jackson. The website is www.profitfromyourpodcast.com.

Let's turn our attention back to *The Pharmacy Memes Podcast*. Co-hosts Jenny and Bill have been through a lot trying to plan and launch their podcast. Podcasting is a lot more work than they expected, and they're starting to think that their original plan of not advertising and just letting people find it is too passive. They're working hard to make this podcast happen, and they want to attract listeners.

Jenny and Bill picked the following five advertising options.

1. Word-of-mouth will be easy. They'll simply spread the word about the podcast to family, friends, co-workers, and colleagues.
2. Episode one would have been a full episode, but Jenny and Bill changed their minds; they're publishing a trailer instead.
3. They are adding a link to the podcast in their personal email signatures.
4. Mandy, the art teacher, is helping Jenny make basic business cards with a QR code from Canva and shipping some to Bill so they can both hand them out at pharmacy gatherings.
5. A call to action didn't make it into the first full episode Jenny and Bill put together, but they plan to include one in future episodes. They will ask their audience to subscribe to/follow the podcast and share it with other pharmacists.

Monetizing the podcast right away is not on Jenny and Bill's radar. But, they can see how getting sponsors down the line might help them pay for their editor, upgrade to a free-standing, domain-based website, cover their monthly media host bill, or buy more equipment next year. First, they want to get year one under their belts so they can see how they like podcasting.

Let's review. We talked about advertising a podcast first and advertising products, services, or businesses on a podcast second. Advertising and monetization are broad topics, and it's up to you to decide how you want to advertise or monetize yours. Podcasting success is not measured only in dollars and cents. Set your expectations before you launch your podcast, and remember your *Why*.

As you plan your podcast, consider how you will advertise your first podcast episode. Pick three options before you launch. As time goes on, research your options and revisit your strategy. Add or subtract elements as your time, money, ambition, and goals allow.

Complete the worksheet for Chapter 11. It will help you plan your podcast.

WORKSHEET FOR CHAPTER 11

Advertising

1. Circle at least three forms of advertising you will use to get the word out about your podcast.

- Social media
- Pod-guesting
- Call to action
- Call to connect
- QR code
- Word of mouth
- YouTube videos
- Blog post
- Podcast trailer

- Newsletter
- Email signature
- Trailer swap
- Episode swap
- Audiograms
- Shorts
- Paid ads
- Dynamic Ad insertion
- Reverse engineering

2. How else will you advertise your podcast?

3. What products, services, or businesses do you plan to advertise on your podcast?

4. Bonus: How will you monetize your podcast?

©2025 Kim Newlove The Pharmacist's Voice®, LLC | www.thepharmacistsvoice.com

CHAPTER 12

Your Recipe

In the first 11 chapters, we covered your *Why*, how to start, time, tools and gear, MP3 files, research, writing, artwork, media hosts, websites, and advertising. You answered the questions on the worksheets, and you have an initial recipe. You may also have ingredients missing from your recipe at this point. In this chapter, we talk about identifying and filling gaps.

Which ingredients are missing from your recipe? When you look at the worksheet for each chapter, what's missing? Did you skip a critical thinking question, assessment, or problem-solving activity? Are you ready to fill in the blank now? Now that you have more information, do any of your answers need to be updated? Find the gaps in your plan and fill them. How do you fill the gaps? Return to what you learned in *Chapter 2: How to Start*. Find resources, listen to people, make decisions, and take action.

Maybe you haven't figured out how to make an MP3 file yet. Pick a DAW. Find resources that will help you learn how to use it. Listen to experts who can fill the gaps in your knowledge or find a Facebook community with knowledgeable members. If you don't want to create your MP3 files, choose an editor, concierge podcasting service, or network. Whatever you do, your goal is to make your first MP3 in 30 days or less.

Chapter 12: Your Recipe

Fill the gaps in your plan until you have everything you need to launch. If you need some help from an expert, consider adding me to your team. I'm available for private coaching. Go to thepharmacistsvoice.com, click the Contact tab, and leave me a message. I'm available by the half-hour or full-hour. I also teach a self-paced podcast planning course for pharmacy professionals. You can find it on kimnewlove.com.

Our friends at *The Pharmacy Memes Podcast* have at least four gaps to fill. First of all, they need to record the trailer for their podcast. They have an episode recorded, but they want to publish the trailer first. Second, they only have two episodes in the works, and they already see the need to create a written production schedule that includes the next 4 episodes. They're scrapping their practice episode with the Mufasa and Simba meme about weekends, but they want to re-record it and use it for episode three. They have memes picked out for episodes four and five. One is about drug shortages. The other is about splitting tablets. Third, they need to create a YouTube channel because they want their episodes syndicated to YouTube. Fourth, they need to double-check their artwork to see if it meets specs. They Googled "podcast artwork specs" before they picked a media host, and they want to make sure everything meets Libsyn's specifications.

Four gaps isn't an overwhelming amount. Jenny and Bill feel good about their podcast plan. They will fill the gaps and launch soon.

If you have filled all your gaps, you're ready to launch! It's an exciting time to be you! Plug the MP3 file, written details, and artwork for episode one into your media host, and schedule the release date and time. Then, get started on episode two. If you want to update the artwork, create social media accounts, or get a co-host down the line, remember that you can change your recipe. Keep an open mind, and just get started! You can do it!

Complete the worksheet for Chapter 12. It will help you identify your gaps and finish planning your podcast.

Happy podcasting!

WORKSHEET FOR CHAPTER 12

Your Recipe

1. What gaps are in your plan?

2. How will you fill those gaps?

3. When will you launch your podcast?

©2025 Kim Newlove The Pharmacist's Voice®, LLC | www.thepharmacistsvoice.com

About the Author

Kim Newlove is an Ohio pharmacist, voice actor, and podcast host. She graduated from The University of Toledo College of Pharmacy with her BS Pharm in 2001. Her clinical experience includes hospital, retail, compounding, and behavioral health.

In 2017, Kim founded The Pharmacist's Voice ®, LLC. Then, she launched *The Pharmacist's Voice® Podcast* in 2019. You can find her business and her first podcast at www.thepharmacistsvoice.com. In 2023, Kim launched her second podcast, *The Perrysburg Podcast*, which you can find at www.perrysburgpodcast.com. As of April 2024, Kim has published more than 300 podcast episodes.

Kim is a go-to pharmacist in audio production. She believes in the transformative power of well-communicated messages, and her mission is to connect with her audience so that they trust her message. Her delivery style is confident and trustworthy.

In her spare time, Kim enjoys spending time with her family, playing *Ticket to Ride Switzerland*, swimming, and riding her BMW C400X motorbike.

NEED HELP STARTING YOUR PODCAST?

You need a plan before you launch.

Complete the worksheets first. Then, book your initial 30-minute appointment with me.

https://calendly.com/kimnewlovevo/30mincoaching

©2025 Kim Newlove The Pharmacist's Voice®, LLC | www.thepharmacistsvoice.com

www.ingramcontent.com/pod-product-compliance
Lightning Source LLC
Chambersburg PA
CBHW042006150426
43194CB00003B/147